Ouassila BEKKAL BRIKCI- BENHABIB

Final year project in biological sciences:

Ouassila BEKKAL BRIKCI- BENHABIB

Final year project in biological sciences:

Preparing a polished manuscript for defence

ScienciaScripts

Imprint

Cover image: www.ingimage.com

This book is a translation from the original published under ISBN 978-620-6-71932-8.

Publisher:
Sciencia Scripts
is a trademark of
Dodo Books Indian Ocean Ltd. and OmniScriptum S.R.L publishing group

120 High Road, East Finchley, London, N2 9ED, United Kingdom
Str. Armeneasca 28/1, office 1, Chisinau MD-2012, Republic of Moldova, Europe
Managing Directors: Ieva Konstantinova, Victoria Ursu
info@omniscriptum.com

Printed at: see last page
ISBN: 978-620-8-35569-2

The aim of this handbook is to guide 'Master 2' students through the process of writing their final project, providing clear and structured guidelines for each essential section. Each stage is designed to ensure [a c[arity, [a coherence and appropriate organisation of academic work in 6io[ogical sciences.

FOREWORD

We live in a world where access to information is easier than ever, but where the quality of that information can sometimes be variable. That's why I've taken on the task of creating a comprehensive and accurate writing guide for Masters students.The aim is simple: to provide every learner with the methodological and motivational strategies to undertake and succeed in academic writing. It is true that lecturers supervising Masters dissertations are often very busy and may find it difficult to find the time to provide detailed training on the rules and recommendations for writing a Masters dissertation. However, there are some good practices that learners can follow to ensure that they produce a quality dissertation.Dear learner, I invite you to read this handbook carefully, which I hope will become a valuable resource for all Masters students in Biological Sciences, providing them with the tools and guidance they need to successfully present and write their dissertation. Remember to maintain a clear, practical and well-structured approach throughout the writing process. By following these tips, you should be able to effectively navigate the sometimes complex landscape of divergent online guides to develop an informative and coherent handbook on writing master's theses.

PREAMBLE

Completing a final year project for a Master's degree in Biological Sciences is an important step in the learner's academic career, enabling them to discover the world of research and acquire reflective, analytical and synthetic skills. The process culminates in the final project being defended in front of a panel of judges, whose assessment is based on the reading of the manuscript, which must be both polished and easy to read.Writing a final year project or dissertation follows a number of rules. This applies not only to the plan, but also to the rules for writing. The latter may seem like a constraint, but the way they are formatted not only makes them easier to read, but also ensures the rigour that is essential if our work is to be reliably disseminated. All dissertations must comply with standards concerning their constituent elements and their form, with the aim of harmonising and facilitating the presentation of dissertations at university. The form of the document must be met if it is to be assessed by members of the jury. This guide is intended for Master 2 students to help them through the writing process. Each section of the dissertation must be written clearly and coherently, following an appropriate structure.This guide can also be used by undergraduates (licence 3) who want to learn about the standards of a university project in order to complete their mini-project.

CONTENTS

1. GENERAL INSTRUCTIONS FOR WRITING A MASTER'S THESIS

To write a dissertation, it's important to follow a few key steps. You have a subject and a supervising teacher, and the test begins. You need to understand the specific expectations and requirements for producing a final year project or dissertation. The second year of the Masters in Biological Sciences ends with the submission of the dissertation and an oral defence. The dissertation is a long-term project, and it is up to the student to plan the work and meet the deadlines.

1.1. Definition of the subject

The supervising teacher will help you to refine your research topic in line with academic requirements. This may include discussions to delimit the subject, formulate clear objectives and define a relevant research question.

1.2. Execution plan for the brief

You are subject to a timetable for completing your dissertation. The plan for completing the dissertation in one semester :
a) Start-up period: Choice of subject with a provisional problem, in conjunction with the supervisor.
b) Period of development of the subject

- Formulation of the problem and the work plan
- Research work
- Preliminary drafting plan

- Definitive formulation of the problem and drafting plan

c) Drafting the brief: A final version of the brief

d) Submission

1.3. Composition of the master's thesis in biological sciences

Structure is the most difficult and important part of writing. You need a clear structure so that readers don't get lost. The structure of the master's thesis / final year project in biological sciences should follow the organisation of an IMRaD scientific article: Introduction, Materials and methods, Results and (and) Discussion **(Swales, 2004).**

This standardised IMRaD outline is intended as a guide to structuring a dissertation in the biological sciences. It provides a clear outline of your scientific work and how the sections are interconnected. The dissertation in biological sciences consists of the following parts:
Cover page (Page de garde)
Signatures (optional) Acknowledgements Contents
List of tables List of figures List of abbreviations
Summaries and keywords (French and English)

1) General introduction

2) Review of the literature (state of the art)

3) Materials and methods

4) Results and discussion

5) General conclusion

6) Bibliography

7) Appendices

1.4.Editorial recommendations

In the biological sciences, writing a master's thesis involves concisely presenting the essential information drawn from several scientific documents relating to the same subject. To write on a scientific research topic, you need to think and propose something new. You need to determine the parameters of your contribution to the construction of knowledge.

1.5.Summarise several scientific articles, chapters and extracts from books

A good summary must be organised and concise, and must convey the information in an unbiased way.

Synthesising a number of scientific articles, book chapters, e-books, website extracts, etc., means presenting new knowledge in a process that combines analysis and synthesis.

The aim is to create a new text that summarises and combines the main ideas from these different sources while preserving their credibility and relevance. In other words, synthesising means :

- Identify scientific articles, book chapters and extracts that deal specifically with your subject. Make sure they are recent and make a significant contribution to your research.

- Extract the key points from the sources, the main arguments and the relevant data.

- Organise information logically and clearly, determining how each source contributes to the overall understanding of the subject.

- Avoid simply reformulating source texts
- Ensure that all the information presented is accurate, citing references correctly.

- Create a coherent text that effectively summarises the knowledge acquired from the various sources, while offering a conclusion that looks back at the initial objectives and offers a new perspective and recommendations.

2. BASIC RULES FOR A SCIENTIFIC STYLE

To communicate well in science, you need to be clear and straightforward, using simple, unambiguous sentences with a precise vocabulary specific to the scientific field **(Roger, 2007).** A dissertation that is free of spelling, grammatical and typographical errors is a guarantee of credibility with the jury. It is therefore important to know how to formulate and use the right terminology to write a successful Master's dissertation.

The scientific style has 4 main characteristics:

- The use of "I" or "we" is not permitted. The passive form must be used. Example: "the objective of this work is..." or "the results show...".
- The use of exact words and short sentences with a vocabulary in line with that of the entire scientific community.
- Use of the present tense (past tense for the methods section).
- It is best to avoid repetition and redundancy.

Generally speaking, the quality of the scientific manuscript reflects the quality of the work carried out during the project.

2.1.Spaces and punctuation marks

When writing a dissertation, the use of spaces and punctuation marks must comply with strict rules in order to articulate ideas. Short sentences avoid the risk of misusing punctuation. In terms of signs and punctuation, the rules must be preceded or followed by either normal or non-breaking spacing (table I).

Table I: Some rules for punctuation marks and symbols

Spaces and punctuation marks		Rules	
		Before	After
Point	.	no space	space
Comma	,	no space	space
Parenthèse	(opening) closing	front space no space in front	no space after Space after
Hook	[opening] closing	front space no front space	no space after space after
Guillemet	"opening closing	front space no front space	no space after space after
Two points	:	unbreakable space	space
Semi-colon	;	unbreakable space	space
Exclamation mark	!	unbreakable space	space
Question mark	?	unbreakable space	space
Points of suspension	...	no space	space
Percentage	%	unbreakable space	space
Mathematical signs	+-x=	Espace	no space
Slash	/	no space	no space
Fraction	2/3	unbreakable space	Espace
Symbols	Kgµm mL °F°C	unbreakable space	Espace
Hyphen	-	no space	no space

At the end of a document or chapter title, punctuation is not used.

2.2. Tables and figures

All figures and tables must be listed, each on a separate sheet. Each figure and table must have a sufficiently self-explanatory title and be numbered in the order in which they are first mentioned in the text.Comparative tables, illustrations and figures can be used to visually organise the information and make it easier to understand.Tables and figures do not have the same role:Tables mainly present numerical and textual data, while figures show visual information (diagrams, lines, curves, charts, etc.).

a) Tables

Tables are of particular interest in the biological sciences because they allow results to be compiled and compared. They often contain figures, percentages, statistics, study results, comparisons and so on.Tables must be numbered in order of appearance, using Roman numerals (Table I, II, III, IV, etc.). They must include a legend so that they can be understood without having to refer to the text of the manuscript. They must therefore be accompanied by a precise informative title.Please note that the title of a table should be placed above it (**table I**) and that tables should not be split over 2 pages, as this would make them difficult to read.

Example: Table taken from a HYGIÈNES journal article (**Bekkal Brikci-Benhabib et al.,** 2015).

Tableau I – Critères des nouveau-nés et prématurés prélevés inclus dans l'étude.

Saxe, poids, âge	Durée d'insertion du cathéter	Antibiothérapie	Référence souche isolée
F - 1,900 kg - 4 jours	4 jours	Ampiciline + Gentamycine	*C. albicans* N103
M - 3,200 kg - 7 jours	3 jours	Ampiciline	*C. albicans* N72
F - 2,600 kg - 4 jours	4 jours	Cefotaxime + Ampiciline + Ciprofloxacine	*C. albicans* N93
M - 3,500 kg - 8 jours	4 jours	Oxaciline +Gentamycine	*C. parapsilosis* N19

b) **Figures**

Figures are used to illustrate concepts, processes, relationships or models visually (**Figure 1**). They can take a variety of forms: diagrams, maps, photographs, histograms, pie charts or graphs... (**Figure 1**).

The representations most often encountered in biological science dissertations are :

- Sectors: presentation of percentages.
- Histograms: static comparison.
- Plots of curves and points: x is the control (explanatory) variable; y is the controlled (explained) variable.
- Photographs, Electron microscopy
- Histological sections, electrophoresis gel...

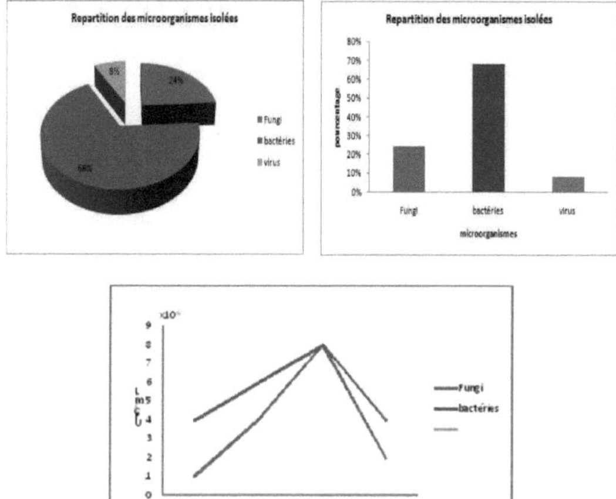

Figure 1: Representations most often found in briefs in the biological sciences

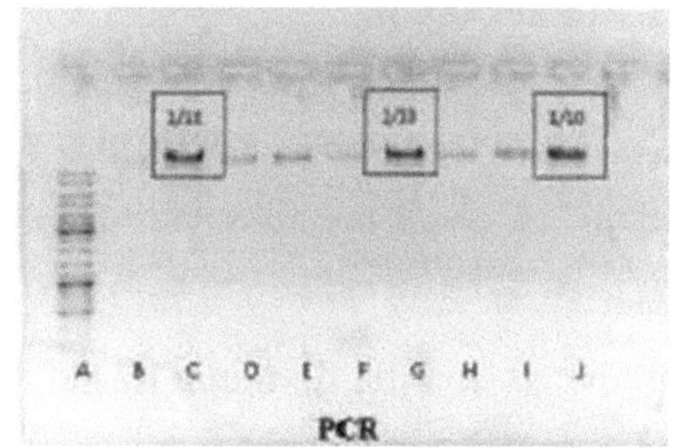

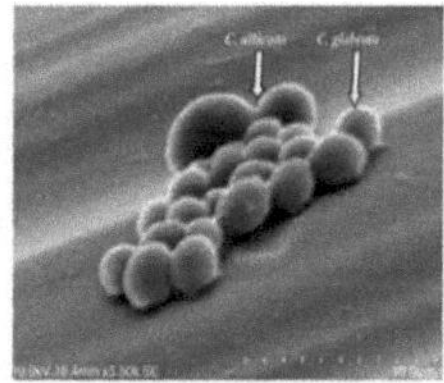

Microscopie électronique à balayage d'un biofilm mixte **(Bekkal Brikci-Benhabib et al., 2021)**

Figures must be referred to in the text and numbered in order of appearance in Arabic numerals (figure 1, 2, 3, etc.). As with tables, the title must be sufficiently informative without having to refer to the text. However, unlike tables, the title is placed below the figure. This allows multiple pieces of information to be summarised in a single representation (**Figure 2**).

Figure 2: Cloud of keywords for a final-year project in biological sciences.

2.3.Abbreviations and acronyms

Unlike the previous lists, this one does not display page numbers. It is in fact an alphabetical list of all the abbreviations and acronyms used throughout the dissertation. Any abbreviation used elsewhere follows (in brackets) its full expression the first time it is cited in the text.

Example: The World Health Organisation (WHO)
The abbreviation O.M.S. may then be used freely throughout the rest of the manuscript. Abbreviations are not recommended in titles, subtitles and summaries. Please note that two Latin abbreviations are permitted and must be used in italics.

- e.g. : exempli gratia, for example ;

- et al: and others.

2.4.Numbers mentioned in the manuscript

Numbers can be written in Arabic numerals, Roman numerals or in full. When a number is followed by a unit of measurement (for example: ml, μ m , km/h), it is expressed in figures. On the other hand, if the number is not followed by a unit of measurement, then numbers from zero to nine are expressed in letters within a text. Numbers greater than ten, on the other hand, are expressed in figures. If the sentence begins with a number, it must be written out in full. Example: Twenty bacterial species belonging to 12 different genera are responsible for... Decimal digits must contain a comma and not a full stop (example: 3.44 and not 3.44). Note that we use a comma in French and a full stop in English. Greek letters, statistical symbols or variables must be written in italics in the text.

2.5.International System of Units

The International System of Units (SI) is the most commonly used in scientific and technical work. It should be noted that the unit symbol must be on the same line as the data value.

Example: Quantity of substance: Mole (mol)

2.6.Biological (or taxonomic) nomenclature

Biological nomenclature, also known as taxonomic nomenclature, is a standardised system for naming and classifying living organisms.

Each species is identified by a binomial name made up of two parts:

- Genre: The generic name, the first letter of which must be capitalised.

- Species: The specific name, written in lower case.

Example: Homo sapiens, where "Homo" is the genus and "sapiens" is the species.

The usual nomenclature for micro-organisms must be used in accordance with international nomenclature rules. Microorganisms should be referred to by their Latin name in italics, with the genus capitalized and the species name lower-case (Escherichia coli or E. coli). These names are italicised in the text. It is possible to use sp. for the species (Clostridium sp.) or spp. which refers to several species (Clostridium spp.).

2.7.Insertions

Terms in foreign languages, particularly Latin, should be underlined or written in italics.

Example: in vitro; in vivo; in silico; et al....

Quotations should be enclosed in inverted commas; if they are long, they should form a separate paragraph and may be in italics or in a smaller font than the rest of the text.

2.8.Formatting

For the layout of the dissertation, a clear and legible font style should be used throughout the document, maintaining consistency in the formatting of headings and sections. It is advisable to use a font such as: Arial size 11 or another common font/size: Calibri 12; Times new roman 12, 1.5 line spacing (For example, this document is written Times new roman 12, 1.5 line spacing). The margins of the text of the dissertation must be equal to 2.5 cm on the left, right, top and bottom.

The text must be justified (aligned from left to right) and spaced at 6 points after each paragraph.

- Bold titles in size 12

- Table and figure legends in size 11.

- The sheets (A4 format) are printed single-sided.

Avoid using fancy fonts (fancy fonts, fancy fonts, etc.) and avoid using colours (use black).With a few exceptions, underlined characters must never be used when writing a scientific manuscript; only italics are permitted.

2.9.Subdivision of sections

The brief must be divided into clearly defined and numbered sections. The sub-sections should be numbered 1.1 (then 1.1.1; 1.1.2, ...), 1.2, etc. The 4ème level must not be exceeded: for example, indicate the paragraph (1.2.1.2.3 ...)With the exception of the title page, all pages must be numbered. Preparatory pages may be numbered (i, ii, iii, etc.). Any sub-section may have a short title. Each title should appear on its own separate line. Please note that the abstract is not included in the section numbering.

2.10. Pagination

Pages must be numbered (the page number must be given in the table of contents). Pagination should be in Arabic numerals (1, 2, 3, etc.). The page number must be at the bottom right of the page. All sections and subsections of the report are numbered from the introduction onwards. Appendices, if any, may also be numbered.

2.11. Volume of the brief

Fifty to sixty pages might be enough to contain all the sections that make up the biological sciences dissertation, excluding appendices. However, some topics may require more pages to cover the complexity of the concepts and data. Note that your department may have specific guidelines regarding the minimum or maximum length of the dissertation. If your work exceeds this number of pages, the jury could criticise you for not having shown sufficient ability to synthesise and for filling in rather than specialising in the objectives of the research you are proposing.

2.12. Ethics in scientific writing

Simple ethical rules apply to writing a dissertation in the biological sciences. All previously published work must be referenced. The citation of a previously published article or borrowed text must be correctly identified, for example by using an italic font and inverted commas. Each element presented in the text must be original. Any conflict of interest, even potential, must be indicated.

1.6 Using artificial intelligence

The learner must not use generative artificial intelligence (AI) such as ChatGPT in the process of writing the dissertation, in order to avoid the abuses generated by this technology (such as Plagiarism).

2.13. Plagiarism

Many learners give in to the temptation to draw inspiration from the work of others when writing their own dissertations, without necessarily realising that they are infringing copyright and incurring sanctions, a warning or permanent exclusion from any higher education institution. A simple way of avoiding plagiarism is to mention a result or idea from someone else, in your own words, giving the reference.

3. GOOD USE OF FRAMING

The Master's supervisor plays a crucial role in the success of your final year project.

a) Before you start

From the start of your collaboration, clarify mutual expectations, including the frequency of meetings, deadlines and preferred methods of communication (email, face-to-face, etc.)

b) During the Project

- Plan regular meetings to discuss the progress of your project and the next steps to be taken. Stick to the agreed times and be ready for each meeting.
- Show initiative and the ability to work independently, while remaining open to advice and guidance from your manager.
- Ask for regular feedback on your work, whether it's on your ideas, your preliminary results, your analyses or the writing of your dissertation.
- You also need to be open to feedback and constructive criticism from your supervisor. Use these comments to improve your work and refine your arguments.
- If you have any difficulties, discuss them openly with your manager so that he or she can offer you advice or alternative solutions.

- Involve your supervisor in preparing your presentation and defence. Practise in front of them.

c) After submission of the brief

After submitting your dissertation, discuss any corrections or improvements to be made based on your supervisor's comments.

Continue to maintain a professional relationship with your supervisor after you have submitted your dissertation. They can be a valuable resource for professional advice and recommendations.

4. ORGANISATION OF THE DISSERTATION

To ensure the consistency of final dissertations, it is necessary to respect the organisation of the sections of which they are composed.

4.1.Cover page

The cover page containing all the information required to identify the brief. It should include : Name of your institution, academic year, specialism, surname and first name of the student, title of the dissertation, names of the supervisor, chair of the jury and examiners and their institutional affiliation, date of the dissertation defence and the institution's logos. With regard to the design of the cover, we recommend using a legible font of an appropriate size, preferably bold for the title of the dissertation. It should be stressed that the dissertation must provide an explicit description of the research subject. The title should summarise the main idea of the manuscript in a simple and original way. It consists of a concise expression and identifies the variables or theoretical problems studied. The recommended length for a headline is 10 to 12 words.

4.2.Summaries and keywords in English and French

The summary is an important section of the Master's thesis, as it summarises the content of the work carried out. For example, the objective of the dissertation, the methodology used, the main results obtained and the conclusions drawn from the process.An English translation of the abstract is compulsory. However, it is important to ensure that the translated information is accurate and consistent. This part of the dissertation is written after the body of the manuscript has been drafted, within a set word limit, usually 250 words (around 20 lines maximum).It should be noted that the summary should be placed before the table of contents and after the acknowledgements. There are no bibliographical references or abbreviations in this section. The abstract should be accompanied by keywords (or combinations of words), which will make it easier to search and index in databases. These terms are limited in number, between 5 and 7 keywords.

Example: Dissertation; Masters; Biological sciences; Methodology.

4.3.Acknowledgements

It is possible to include a dedication before the acknowledgements page. This space is optional for the brief.Acknowledgements are intended for the members of the jury, supervisor/co-supervisor, chairman of the jury, examiners, and any person or institution that made a contribution during the research and preparation of the final year project may be mentioned in this section. These acknowledgements must be written in correct terms.

4.4.Contents

The table of contents must be cross-referenced with the corresponding page number. It contains all the parts and sub-parts with their respective numbers as they appear in the manuscript, so that you can find your way around it quickly. Pages for dedications, acknowledgements, abstracts, etc. are not included. Pagination begins with the introduction. This means that the first numbered page will be the first page of the introduction. The cover page (or title page) and preliminary pages such as the abstract, summary, lists of tables/figures, etc., are often numbered in Roman numerals (I, II, III, etc.). However, these pages do not show a number on the page itself (**Figure 3**). The table of contents is single-spaced, except for the section headings, which are single-spaced, without bold, italics or underlining. A table of contents can be generated automatically using word-processing software (this is strongly recommended).

Contents

Figure 3: Table of contents

The summary of the brief, detailing the sections with numbered cross-references.

4.5.General introduction

The purpose of the introduction is to set the research in a more general context. It should take the reader from a general problem to a research question, justifying the relevance of this question. The facts and opinions put forward in the introduction must be supported by bibliographical references (**Lesage, 2013**). The introduction provides information on the subject of the dissertation, as well as the general context in which the subject is studied. It sets out the aims and objectives of the dissertation.A good introduction should motivate the reader to take an interest in the manuscript as a whole. So, as well as being well written, the introductory text should be an attractive invitation to continue reading.The problem statement is in the form of a precise question (and the specific research questions around it) and the choice of key concepts delimiting the literature review. It will evolve as the dissertation progresses. Generally speaking, this part will not exceed two pages in a Master's dissertation. As far as verb tenses are concerned, the use of the present tense for admitted facts is recommended, and the use of the past tense for quoting authors or stating the purpose of the work.

4.6. Literature review / State of the art

The aim of the literature review (state of the art) is to encourage you to structure a coherent, well-documented review of existing knowledge on a specific subject.

The key stages in producing an effective literature review or state of the art :

- **Defining the subject**: Define the subject of your research in order to target the relevant sources. Identify the concepts and areas that are important for your study.
- **Gathering sources**: Search for scientific articles, book chapters, theses and other relevant documents in academic databases, online libraries and other specialist sources.
- **Critique of sources**: Evaluate the quality and relevance of sources according to criteria such as the reputation of newspapers or publishing houses, the methodology used and the validity of the results presented.
- **Analysis and organisation**: Identify emerging trends, divergences and convergences in the literature.
- **Synthesis**: Integrate the information extracted from the various sources by synthesising it into a state-of-the-art report.
- **Citations and references**: Be sure to properly cite all sources used according to the appropriate academic citation standards (APA, MLA, etc.). This ensures the intellectual integrity of your work.
- **Updating**: The literature review should be updated regularly to include the most recent research and to reflect the evolution of knowledge in the field.

In general, this section should not exceed 20 pages, so it should not be presented as a chapter but only as a heading. The present tense should be used to describe or present the scientific context (data from the literature). Even simpler, what has already been published should be described in the present tense.

4.7.Materials and methods

The "Materials and methods" section of a Masters dissertation is crucial because it describes in detail how you conducted your study. This part must be descriptive, precise and sufficiently detailed. In general, it will deal with methods that allow traceability, the possibility of reproducing experiments and enable the dissertation to be judged objectively and clearly.

The relevance of writing this section well is :

a) Description of equipment

- Describe any specific equipment you used to collect data, e.g. measuring devices, software, experimental apparatus, etc.
- List all the materials, reagents, chemicals and other consumables used in your research.

b) Description of Methods

In this section you should explain the general framework of your study, including the type of study (experimental, observational, etc.). The methods you followed should be described step by step, including sufficient detail for someone else to be able to replicate your experiment.Data collection, including sampling methods, inclusion/exclusion criteria, questionnaires, measures adopted, statistical or analytical techniques you have used to analyse your data and any other relevant considerations are necessary in drafting this "Methods" section.

Examplel: Description and origin of microorganisms (wild or reference), plants (or animals) [genus, species, strain, cultivar...].

Example2: give the exact reference of the software used (Statistica).

It should be noted that when describing or briefly summarising standard methods and techniques, it is imperative to give the source bibliographic reference which describes them precisely. If one or more modifications are made to the methods section, it is important to give the source bibliographic reference with a description of the modification made.

c) Limitations of the study

Identify and discuss any potential limitations in your methodology or equipment that may have influenced your results.

4.8.Results and discussion

The 'Results and Discussion' section should be well structured and demonstrate your thorough understanding of the subject and its implications for your Master's thesis. This part can take two forms: either "Results and discussion" are in the same section, or as two separate sections (10. Results
- 11. Discussion).

a) Results

The results must be presented in a clear and organised manner, using tables, graphs, figures and textual descriptions. The "Results" must not contain any methodological elements, discussions or bibliographical references. This section presents the results of your research. This part should be written in the past tense. Sometimes the present tense is used when describing a table or figure directly, and the past tense when referring to the analysis that has been carried out and the results. Typically, we begin by presenting the data relating to the sample and the quality of the results: response rate, description of the sample, comparison with the reference population (representativeness). We then present the results that are useful for answering the research question.Both positive and negative results should be indicated. Invalidating a hypothesis is just as important as validating it and can be discussed in the following "discussion" section.

b) Discussion

It allows the facts and results obtained to be sorted and interpreted. By rigorously analysing the data and relating them to each other, we can highlight the main ideas. In this part of the dissertation, comparisons must be made with the literature. In fact, the results obtained must also be compared with the existing literature, whether there is agreement or disagreement and any possible contradictions. You should also discuss the limitations of your study that may have influenced your results and their interpretation. These limitations serve to temper the significance of the results by pointing out that, as in all research, there are a certain number of limits, biases or unknowns to be taken into account. In the discussion of the results, the author's personal opinions are written in the present tense. The future conditional tense is used to describe a work project. Precision contributes to the clarity of the writing and therefore to its quality.

4.9. General conclusion

The general conclusion of a master's thesis provides an assessment of the subject, showing its implications for an understanding of the subject studied and setting out certain recommendations. It is often redundant with the results and discussion section.

By writing a careful and well-structured overall conclusion, you will ensure a coherent and convincing finish to your Master's thesis, reinforcing the impact of your research and the contribution it makes to your field of study.

Providing a solid conclusion to your Master's thesis paves the way for further research and practical applications of your work. The conclusion should be written in the present tense. By adhering to these rules, you not only ensure that your brief is readable, but also that it is credible. A well-structured, well-written document inspires confidence and credibility in those who read it.

4.10.Resources and references used

The purpose of the bibliography is to support the argument of your work and to show the extent of the bibliographical research carried out. It is essential to include bibliographical references to lend credibility to your work by drawing on validated information. References are a section in their own right and should be placed after the conclusion. Particular care must be taken when writing bibliographical references. This writing must strictly comply with a few simple but imperative criteria:

- be strictly accurate;

- comply with the required formatand becomplete (authors, title, journal or publisher, reference to year, volume, issue and pages).

- Each reference must be easy for the reader to find (and therefore check).

- References can be listed alphabetically by author's name.

- Bibliographical references must be cited in the text before any punctuation.

- In the text: The name of the author(s) appears in brackets in the text, followed by the date of publication.

a) **One or two authors**: The authors' names are always displayed, separated by the coordinator "and" (with no comma before the coordinator).

The document plays a grammatical role in the sentence: According to **Rimondini (2022)**, bacterial biofilm...

We disagree with **Mukherjee and Bassler (2019)**, who show ...

b) **More than two authors**: In principle, only the name of the first author is displayed, followed by "et al. (in italics, with a full stop), then the date.

- The document has no grammatical role in the sentence: **(Stoodley et al., 2022) (Mukherjee and Bassler, 2019).**

- The document plays a grammatical role in the sentence: **Stoodley et al (2022)** explain...

- When the same author (or a group of the same authors) has published several articles in the same year, the different documents are classified in alphabetical order of title, by adding a letter to the date of publication: **(Ramage, 2006a), (Ramage, 2006b).**

- If several references are used, they are arranged alphabetically by author, then by date, and separated by a semicolon; the names of identical authors are not repeated **(Johnston, 2004; Ramage, 2001, 2006; Sauer et al., 2022; Stewart et al., 2022).**

4.10.1. Types of sources

a) **Chapters, guides and books**

- Beaud, M., & Gravier, M. (2006). L'Art de la thèse. Comment préparer et rédiger une thèse de doctorat, un mémoire de maîtrise recherche ou tout autre travail universitaire à l'ère du Net.

- Swales, J. M., & Feak, C. B. (2004). Academic writing for graduate students: Essential tasks and skills (Vol. 1). Ann Arbor, MI: University of Michigan Press.

- Bekkal Brikci-Benhabib, O, (2021). Bioadhesion: development in biofilm. KS Omniscriptum Publishing , ISBN-13: 9786203427264

b) **Magazine articles**

- ***Lesage, F.-X. (2013). Writing a scientific article. Archives des Maladies Professionnelles et de l'Environnement, 74(6), 649-654. https://doi.org/10.1016/j.admp.2013.09.006***
- ***Smith, R. (2000). Quality improvement reports: a new kind of article: They should allow authors to describe improvement projects so others can learn. BMJ, 321(7274), 1428. https://doi.org/10.1136/bmj.321.7274.1428***

c) **Institutional guides and online resources**

- ***Abdellatif, B. D. (2023). Documentary research and dissertation design. 2éme Année Master (Doctoral dissertation).***

d) **Articles from reliable blogs and websites**

- ***The Thesis Whisperer (2023). "Top 10 Tips for Master's Thesis Writing." Retrieved from https://thesiswhisperer.com/top-10-tips-masters-thesis-writing/.***
- ***Guide to writing an internship report in applied biochemistry http://elearning.univ-temouchent.edu.dz/course/view.php?id=582***

Various standards exist, but in this guide we will use the APA standard, which is currently the world reference. The APA (American Psychological Association) bibliographic style was used in the writing of the final year project in biological sciences.

[American Psychological Association (2003). APA style tips. https://apastyle.apa.org/learn/faqs

[2] Aronson E. Methods of research in social psychology, New York, McGraw-Hill, 1990.

[Provost, M. A., Alain, M., Leroux, Y., & Lussier, Y. (2002). Normes de présentation d'un travail de recherche. Trois-Rivières: Les Éditions SMG.

[Yousfi, B. (2021). Une démarche de (co) construction des compétences en contexte universitaire algérien: le cas de l'apprentissage des normes de rédaction académique. Synergies Algérie, (29), 265-281.

4.10.2. Choice of bibliographic management software

Several bibliographic management software packages are available, each with its own advantages and specific features. To choose the software best suited to your needs in the biological sciences for writing master's theses, consider aspects such as the ease of use of the interface, the ability to manage PDFs and bibliographic data specific to your field, and support for citation styles.

- **Zotero** is free, open-source software that is widely used for managing bibliographic references. It supports many citation styles in the sciences, including those used in biological publications.
- **Mendeley** combines bibliographic management with academic social networking features. It allows you to store and organise documents, as well as generate citations and bibliographies. This software is particularly popular in scientific fields for its ability to manage collections of PDF documents and facilitate collaboration.
- **EndNote** is a paying software package widely used in the academic world for managing references. It is compatible with many citation styles, including those specific to publications in the biological sciences.

4.11. Appendices

The appendices include any additional technical documents or methods that would not otherwise be included in the main text, and which could make it unnecessarily cumbersome. The appendices are announced in the table of contents, then in the body of the text (**Appendix 1**).They are numbered consecutively and their pagination must follow the pagination of the main text.

5. ASSESSMENT CRITERIA FOR THE DISSERTATION

The criteria for assessing a Master's thesis may vary from one institution to another. However, there are generally common elements that examiners and juries use to assess the quality of a master's thesis.

- Relevance of the subject: The originality of the subject in relation to the field of study, as well as the clarity and relevance of the issues and objectives of the dissertation.
- Theoretical framework: The quality of the bibliographic synthesis and the demonstration of an in-depth understanding of the context in which the research theme takes place.
- Research methodology: The research methods chosen and used to answer the research question.
- Analysis and interpretation of results: The ability to critically analyse the data collected and to interpret the results in a manner consistent with the theoretical framework and to draw relevant conclusions.
- Structure of the dissertation: The clarity and coherence of the structure of the dissertation, including the presentation of the sections (parts). The adequacy and quality of the references used to support the arguments put forward in the dissertation and compliance with appropriate citation standards (APA).
- Oral presentation: The quality of the oral presentation of the dissertation at the defence, including the ability to answer questions and defend the results obtained.

6. ORAL PRESENTATION / DEFENCE

The defence is an oral examination during which the student presents his/her work and answers questions from a jury. The jury is made up of at least two people, including the supervisor of the dissertation and a teacher-researcher, who has previously examined the dissertation and prepared comments and questions. A member of the professional world, such as a socio-economic partner, may be invited to take part in the examination. The jury is chaired by a teacher-researcher. It is advisable to plan the defence in agreement with the supervisor, so as to aim for the June session rather than the September one. It is important not to exceed the allotted time, which is 15 to 20 minutes maximum.

Tips for a successful dissertation

- Benefit from the support of your teacher and be open to feedback. Learn constructively to improve the quality of your work.

- Be prepared for the viva by anticipating potential questions.
- Be clear and concise in your presentation and your answers.

By following these steps, you will increase your chances of producing a high-quality dissertation and successfully defending it.

You need to do some research to back up your arguments, and make sure you cite your bibliographical sources correctly. Organise your work logically, including a clear introduction, development and conclusion. By following these steps and taking advantage of the guidance provided by your supervising teacher, you will increase your chances of producing a high-quality dissertation and passing your viva successfully.

7. POST-SUPPORT PROCESS

The final dissertation post-sitting process enables the dissertation to be refined so that it is as accurate and complete a reflection of the research carried out as possible, thus guaranteeing its formal acceptance by the department and, consequently, the award of the Master's degree.

a) **Correction and Revision**

After the defence, you may have to make corrections or revisions to your dissertation based on the comments and suggestions received from the members of the jury during the defence. Any necessary changes must be made to the final document before it is submitted.

b) **Assessment and validation**

Once submitted, your dissertation may be checked one last time to ensure that it meets the required academic criteria.

c) **Administrative formalities**

After the correction phase, the finalisation of the dissertation concludes with the administrative formalities involved in submitting your dissertation, such as signing the submission documents.

d) **Archiving or publication**

Your dissertation may be archived in the university library or published in an academic database.

8. REFERENCES

1. Beaud, M., & Gravier, M. (2006). L'Art de la Thèse. How to prepare and write a doctoral thesis, a research master's dissertation or any other academic work in the age of the Net.

2. Bekkal Brikci -Benhabib, O., Boucherit-Otmani, Z., Boucherit, K., & Djediat, C. (2021). Interaction in a dual-species biofilm of Candida albicans and Candida glabrata co-isolated from intravascular catheter. Microbial Pathogenesis, 152, 104613.

3. d'Ortho, M. P. (2012). Second year of Master: "Biology, physiology, cardiovascular and respiratory pharmacology", year 2010-2011: presentation. Journal of Respiratory Diseases, 29(7), 924-925.

4. Lesage, F.-X. (2013). Writing a scientific article. Archives des Maladies Professionnelles et de l'Environnement, 74(6), 649-654. doi:10.1016/j.admp.2013.09.006

5. Rogers, S. M. (2007). Mastering scientific and medical writing. Springer- Verlag Berlin Heidelberg.

6. Serra, Éric (2013). Notes de lecture. Médecine Palliative : Soins de Support - Accompagnement - Éthique, 12(3), 162-. doi:10.1016/j.medpal.2013.03.002

7. Smith, R. (2000). Quality improvement reports: a new kind of article: They should allow authors to describe improvement projects so others can learn. BMJ, 321(7274), 1428.

8. Steriu, L. (2020). The practice of secondary citation in the master thesis. Le point de vue normatif. Mélanges francophones, 15(18).

9. Zouhry, I., & Sannier, A. (2023, January). Le Master 2 Recherche pour les internes d'anatomie pathologique: pourquoi? Why and how? In Annales de Pathologie (Vol. 43, No. 1, pp. 25-28). Elsevier Masson.

ANNEXES

Annex 1: Anti-plagiarism

Declaration of honour form

Commitment to respecting the rules of scientific authenticity in the preparation of a research project

I, the undersigned,

Student, Mr or Miss.. Registered at

....

And in charge of preparing a Master 2 thesis in ....

Title: ....

I declare on my honour that I undertake to comply with the scientific and methodological rules and standards of professional ethics and academic authenticity required in the preparation of the research project in question.

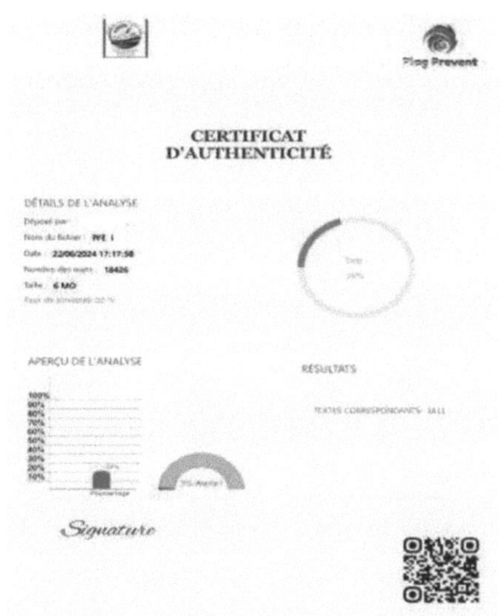

ABSTRACT

The success of a 'Master 2 in 6io[ogical sciences' dissertation is not just a matter of the quantity of scientific information presented, but above all the qua[ity of the structuring, ['ana[ysis and ['interpretation' of this information. <n effet, une structure c[aire et organisée aide [es[ecteurs à comprendre faci[ement [e contenu du mémoire. Cette[a permet de présenter [es informations de manière [ogique, en suivant une progression nature[[e du pro6[ème de recùercùe au.x conc[usions. <By following the guidelines in this guide, you can produce a dissertation that not only meets academic requirements but also reflects your ability to carry out rigorous scientific work in the 6io[ogical sciences.

Printed by Books on Demand GmbH, Norderstedt / Germany